Teaching Resources

Grade 4

Harcourt Brace & Company

Orlando • Atlanta • Austin • Boston • San Francisco • Chicago • Dallas • New York • Toronto • London

http://www.hbschool.com

ISBN 0-15-311119-4

7 8 9 10 085 2001

CONTENTS

GEOMETRY

DATA, PROBABILITY, AND GRAPHING

For use with specific activities

PUPIL'S EDITION LESSONS

TEACHER'S EDITION PRACTICE GAMES

► TEACHER'S EDITION, TAB B, PRACTICE ACTIVITIES

► LEARNING CENTER CARDS

Name _____

Problem Solving

Understand

1. Retell the problem in your own words. _____

_____

_____

_____

2. List the information given. _____

_____

3. Restate the question as a fill-in-the-blank sentence. _____

_____

Plan

4. List one or more problem-solving strategies that you can use. _____

_____

5. Predict what your answer will be. _____

_____

Solve

6. Show how you solved the problem. _____

_____

_____

_____

7. Write your answer in a complete sentence. _____

_____

Look Back

8. Tell how you know your answer is reasonable. _____

_____

9. Describe another way you could have solved the problem. _____

_____

Problem-Solving Think Along

Understand

1. What is the problem about?

2. What information is given in the problem?

3. What is the question?

Plan

4. What problem-solving strategies might I try to help me solve the problem?

5. About what do I think my answer will be?

Solve

6. How can I solve the problem?

7. How can I state my answer in a complete sentence?

Look Back

8. How do I know whether my answer is reasonable?

9. How else might I have solved this problem?

4	0
5	1
6	2
7	3

12	8
13	9
14	10
15	11

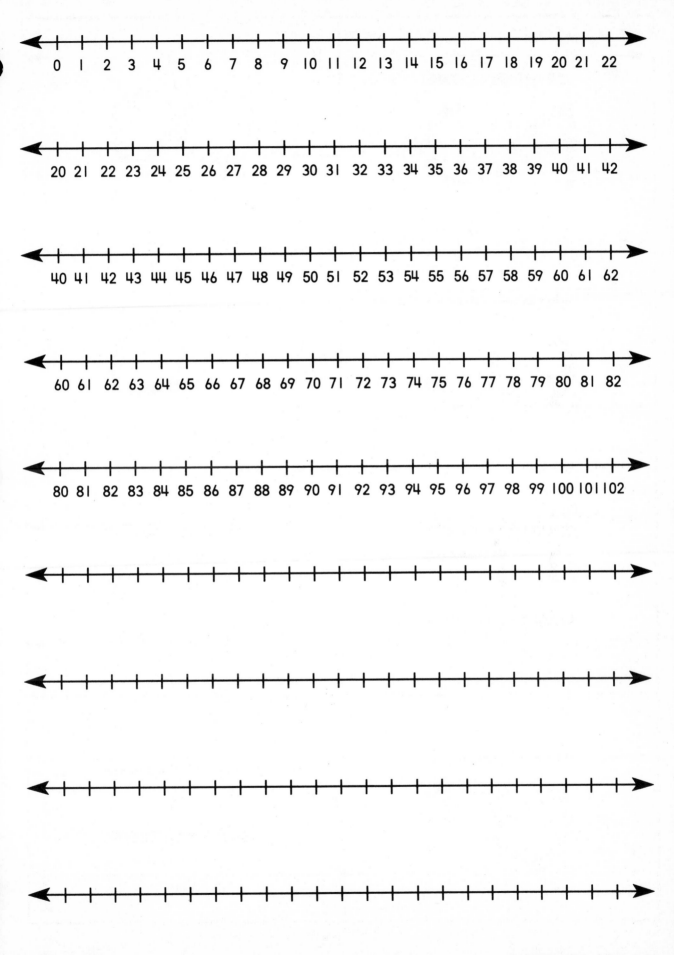

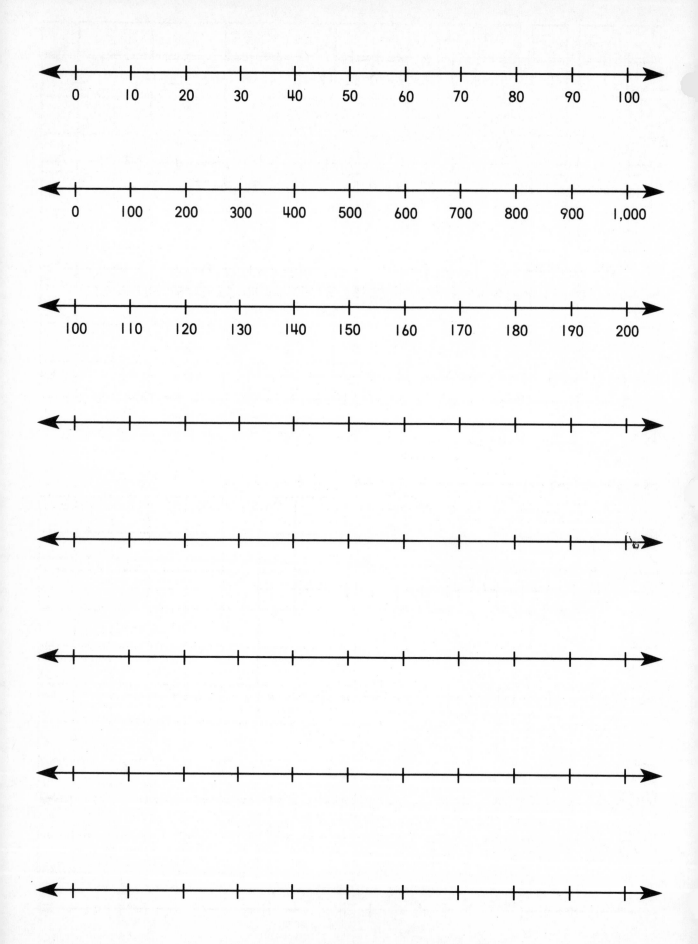

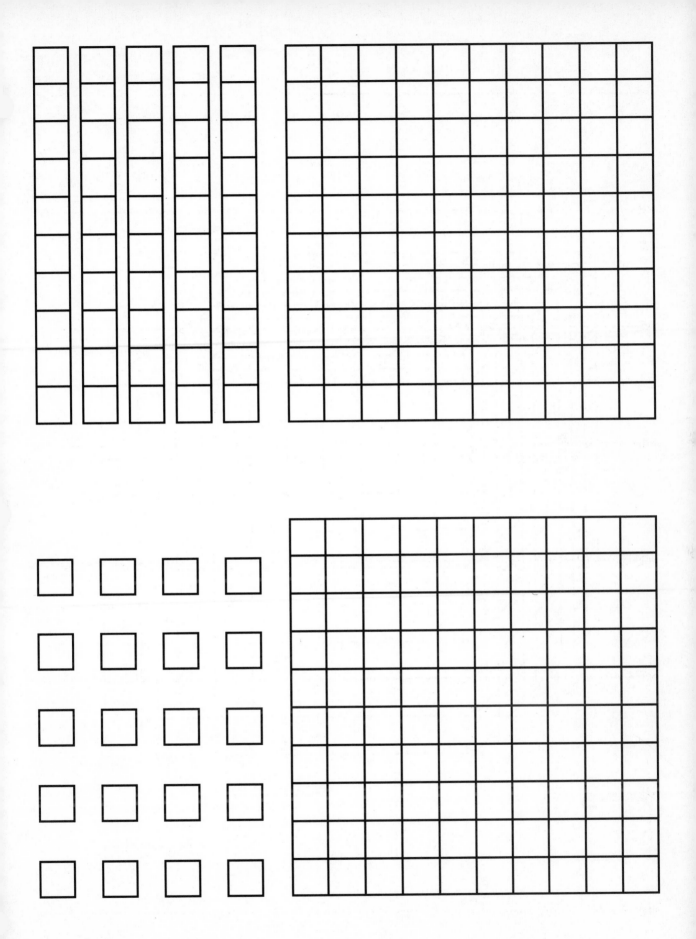

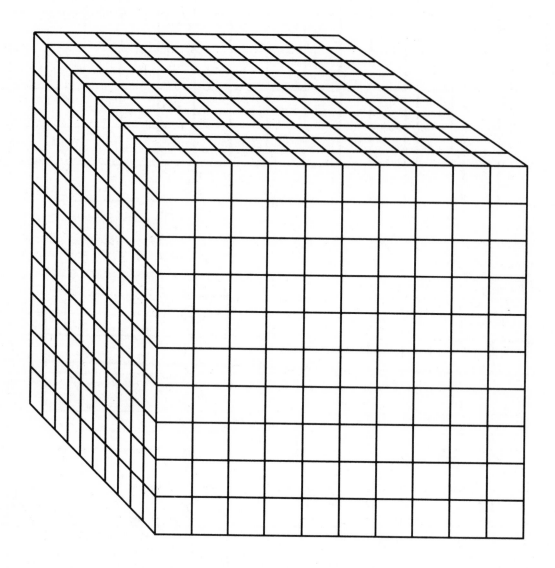

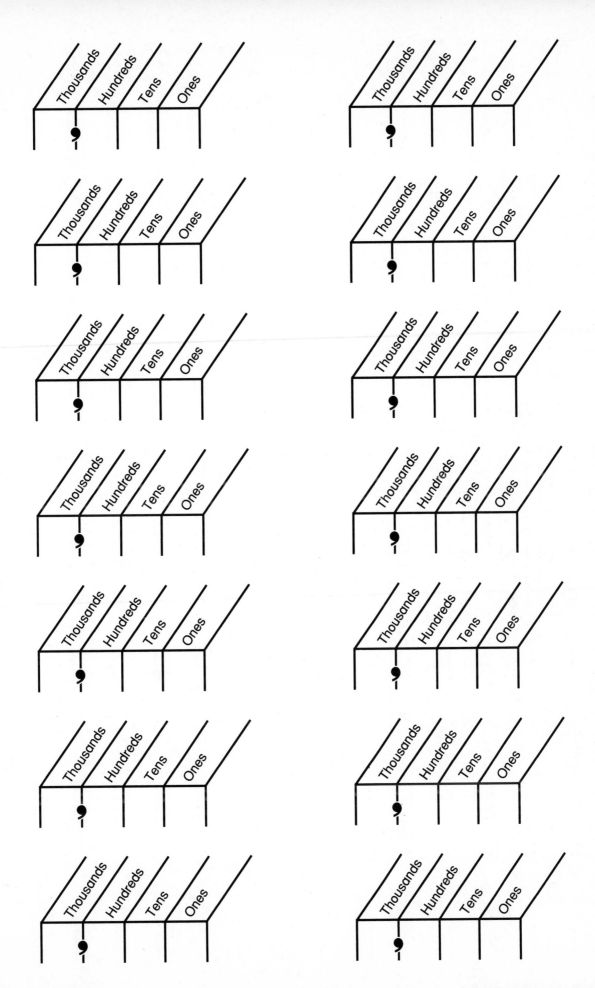

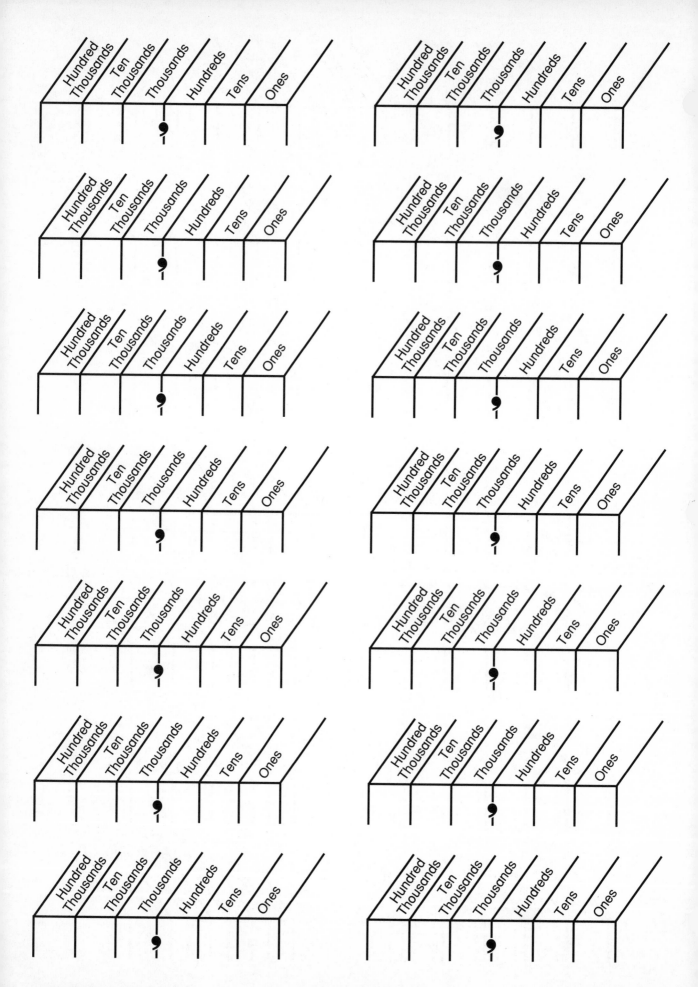

Place-Value Charts

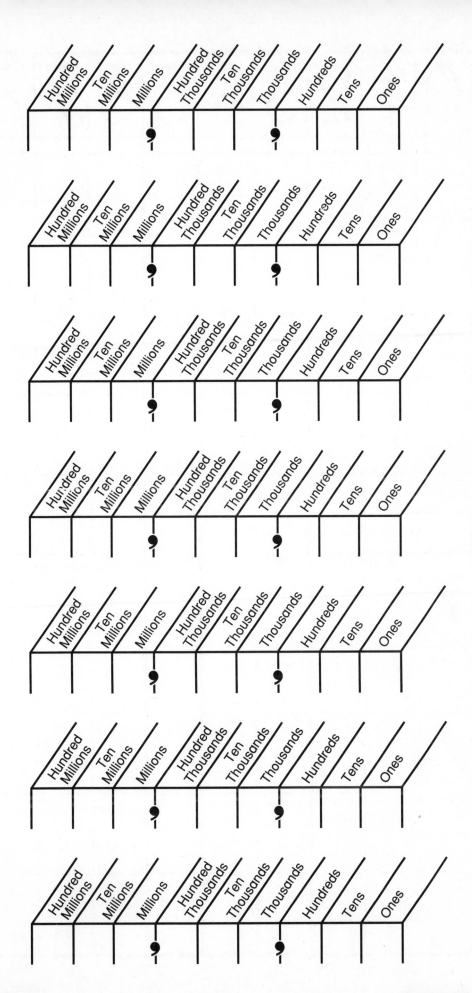

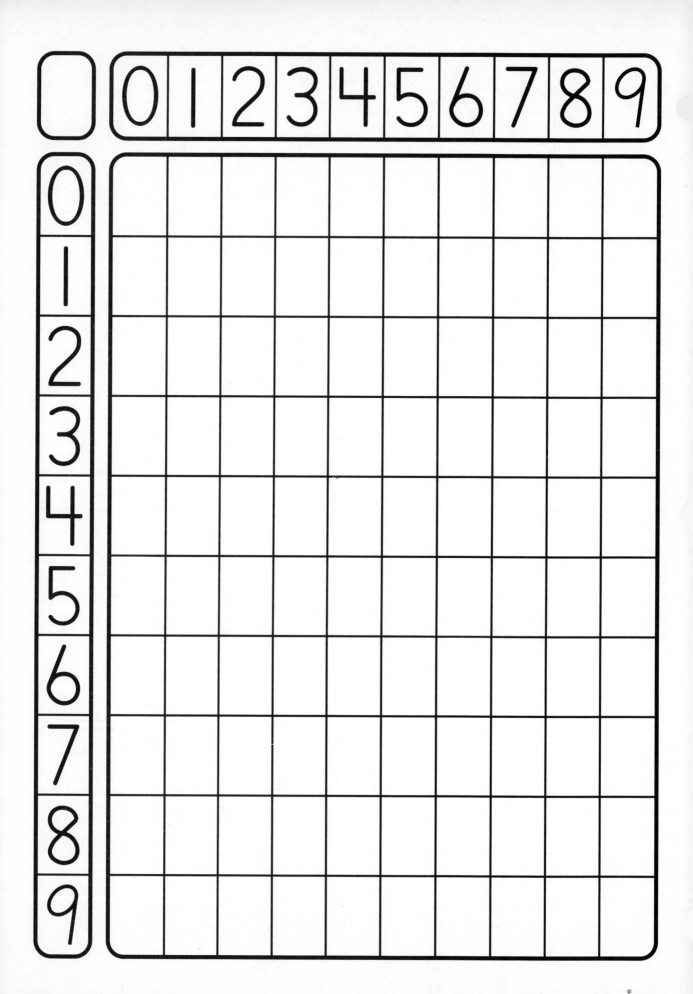

Addition/Multiplication Table

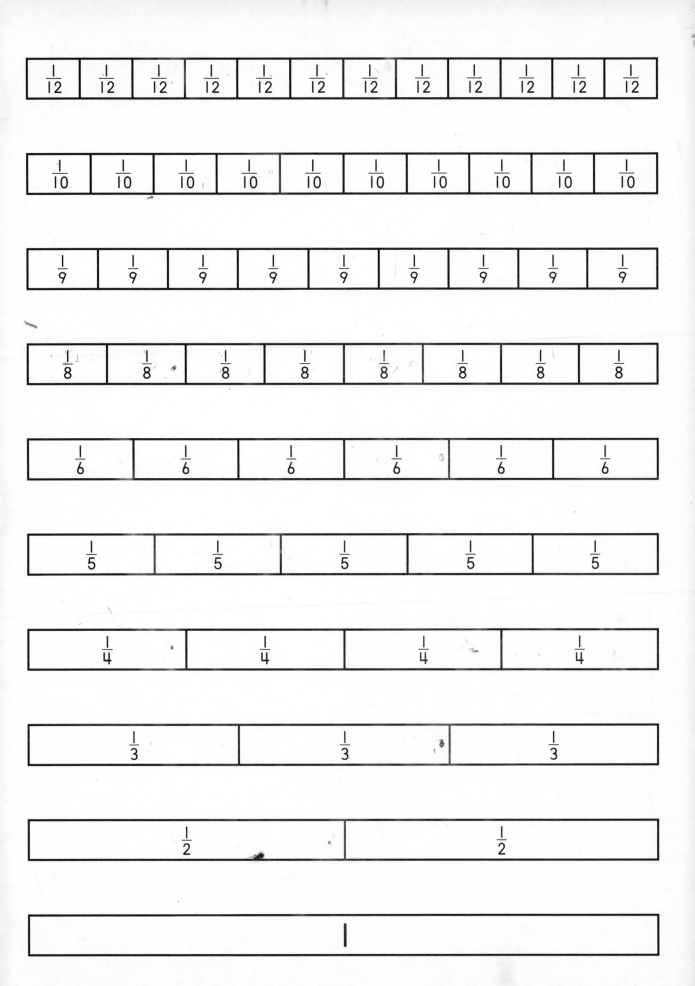

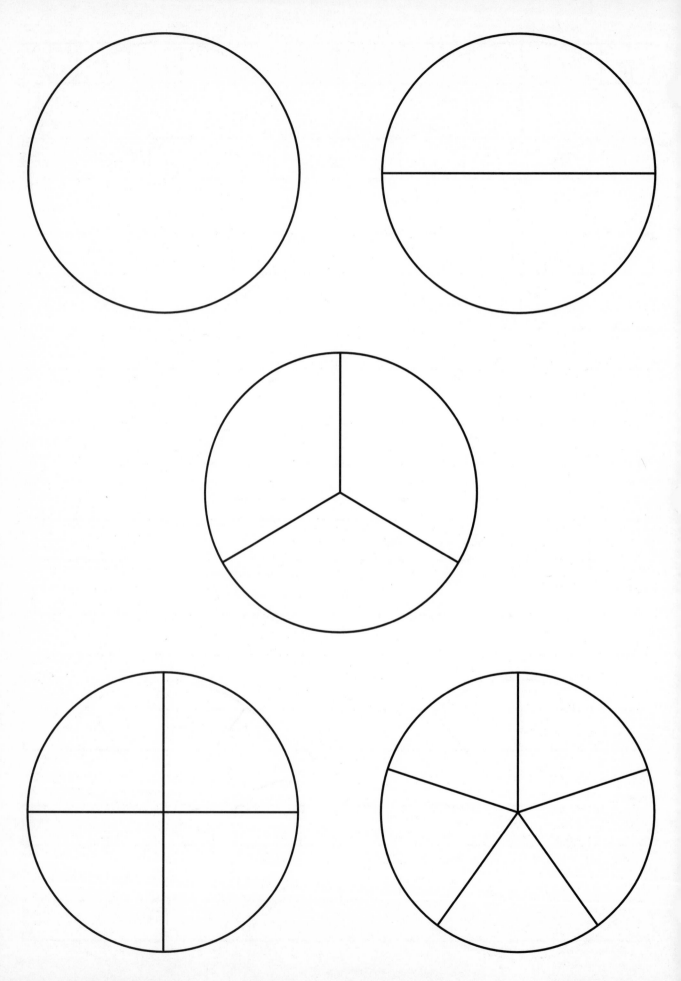

Fraction Circles

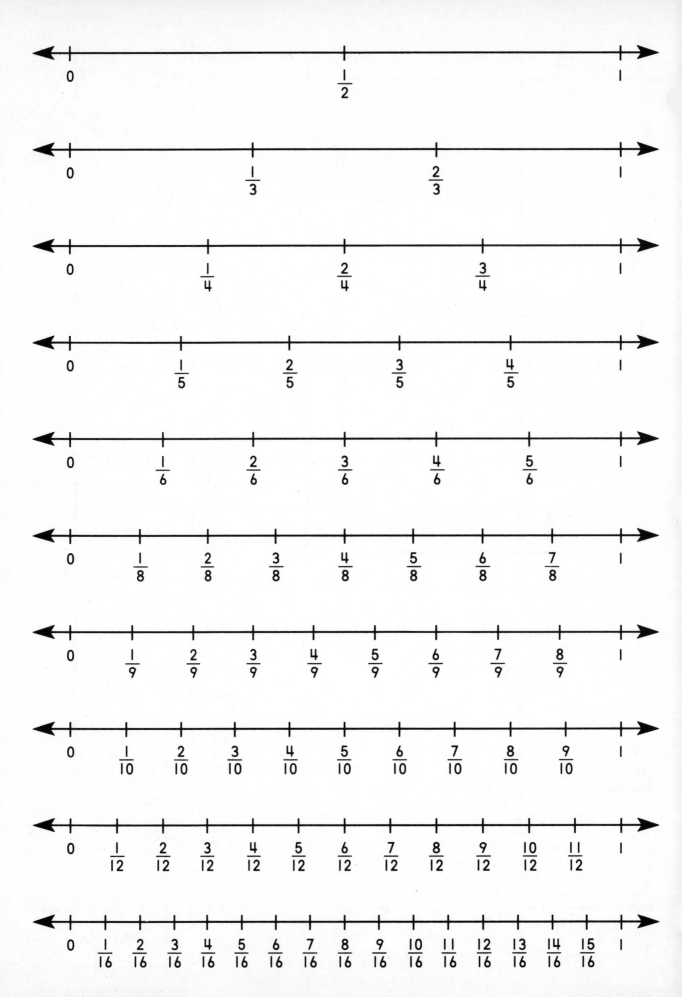

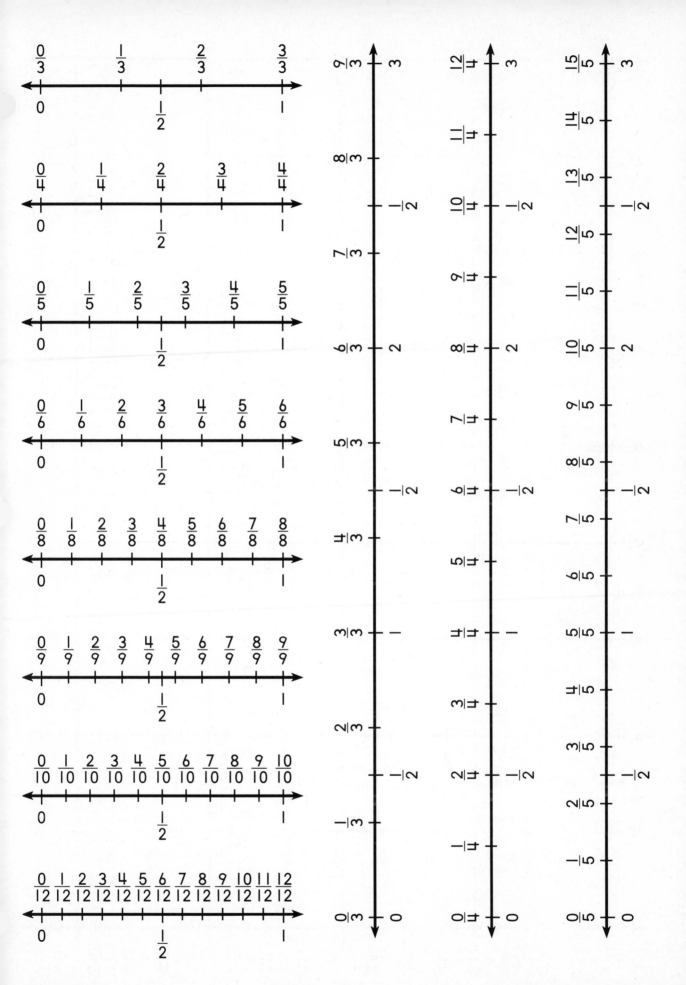

Number Lines

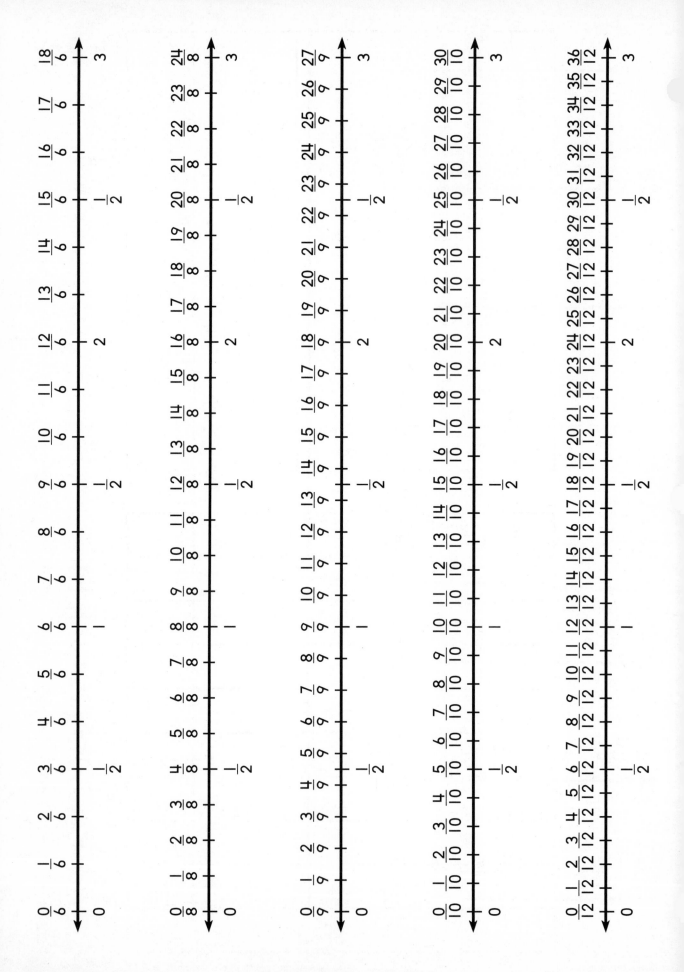

Decimal Squares

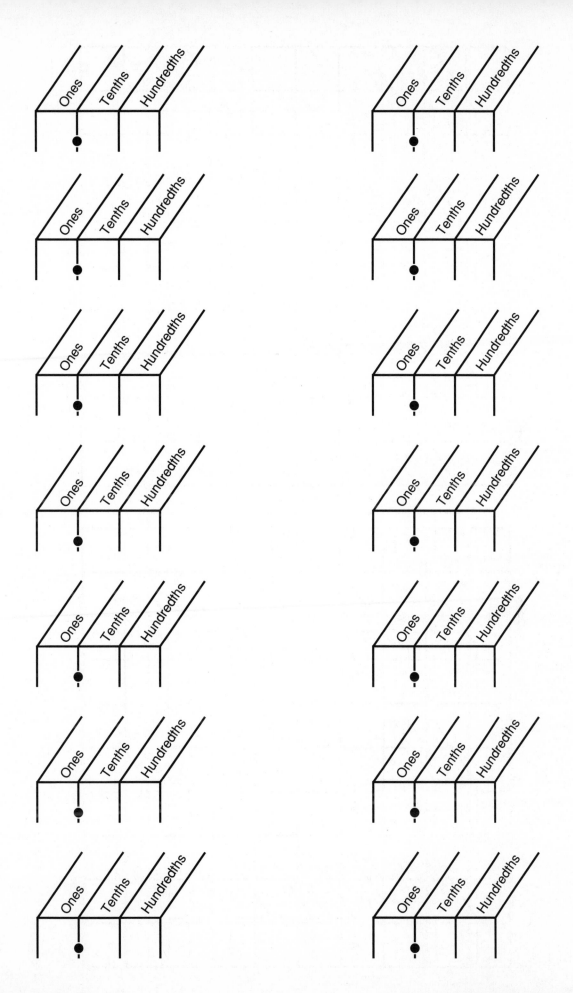

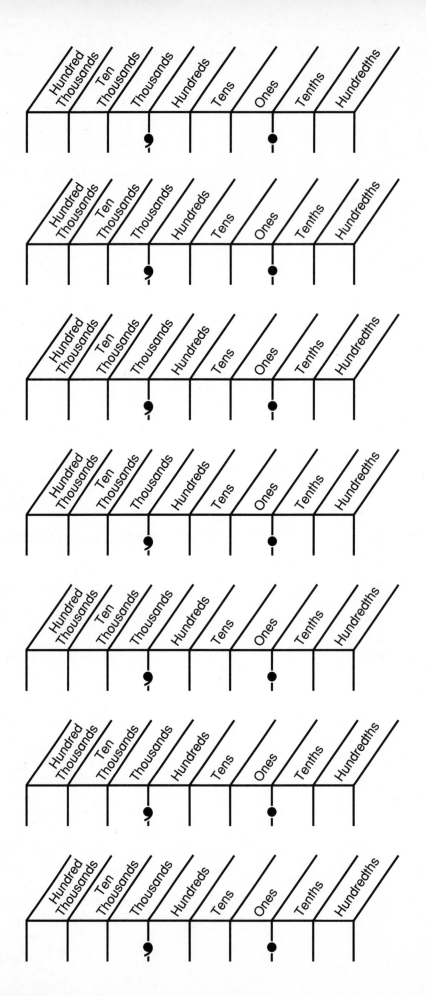

Place-Value Charts

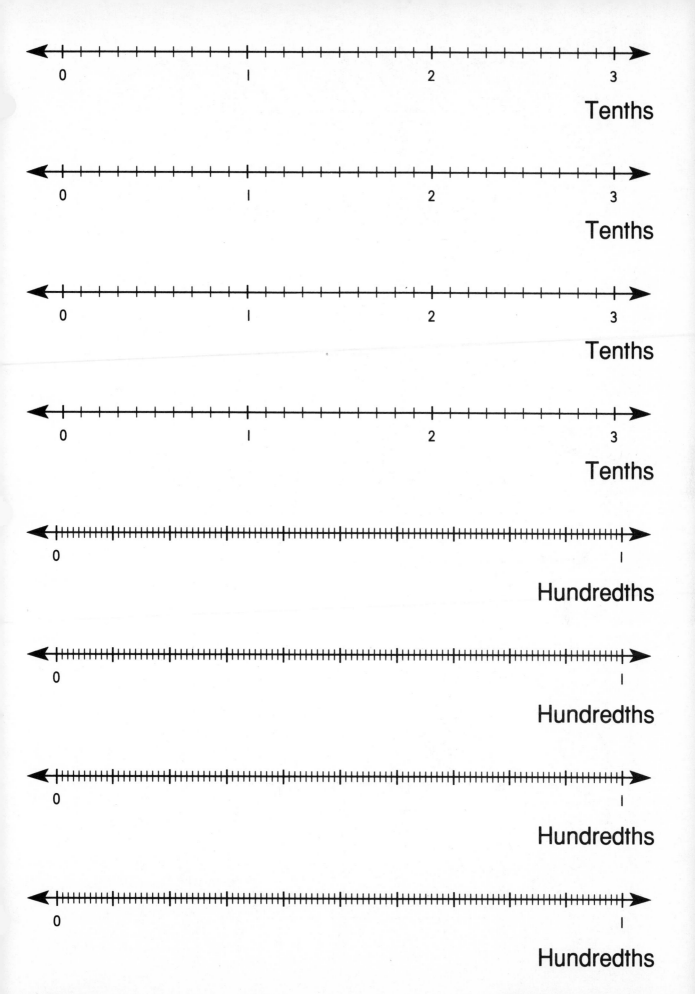

Tenths

Tenths

Tenths

Tenths

Hundredths

Hundredths

Hundredths

Hundredths

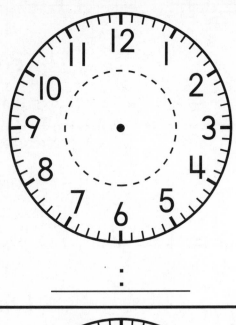

:

_____

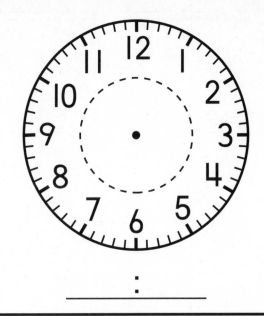

:

_____

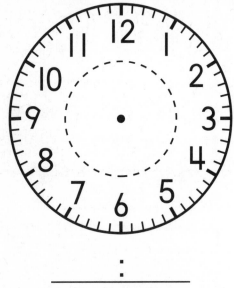

:

_____

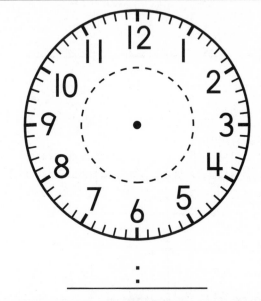

:

_____

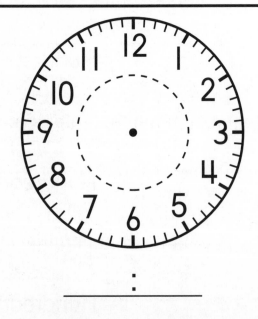

:

_____

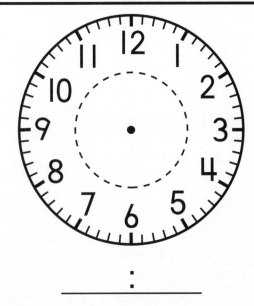

:

_____

Analog Clockfaces

Coins and Bills

Sunday	Monday	Tuesday	Wednesday	Thursday	Friday	Saturday

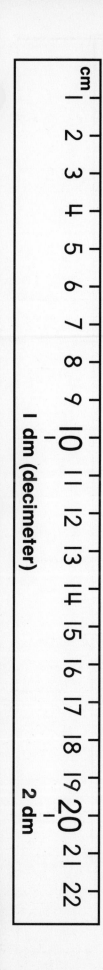

inches

1 2 3 4 5 6 7 8 9

inches

1 2 3 4 5 6 7 8 9

cm

1 2 3 4 5 6 7 8 9 10 11 12 13 14 15 16 17 18 19 20 21 22

1 dm (decimeter)   2 dm

cm

1 2 3 4 5 6 7 8 9 10 11 12 13 14 15 16 17 18 19 20 21 22

1 dm (decimeter)   2 dm

Rulers

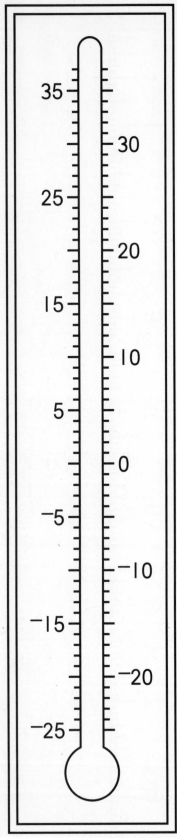

Celsius

____ °C

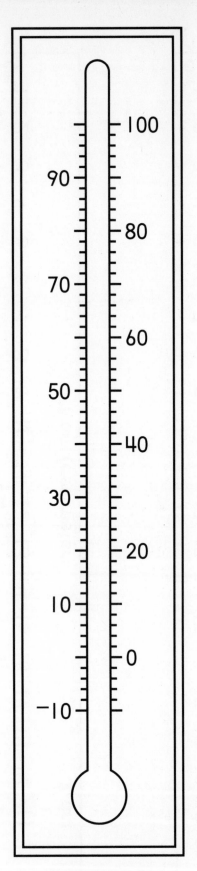

Fahrenheit

____ °F

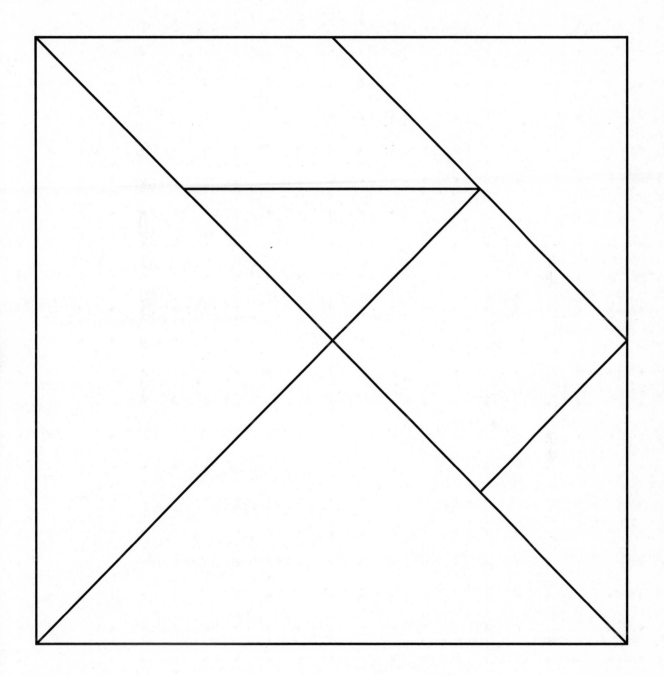

Tangram Pattern

Dot Paper

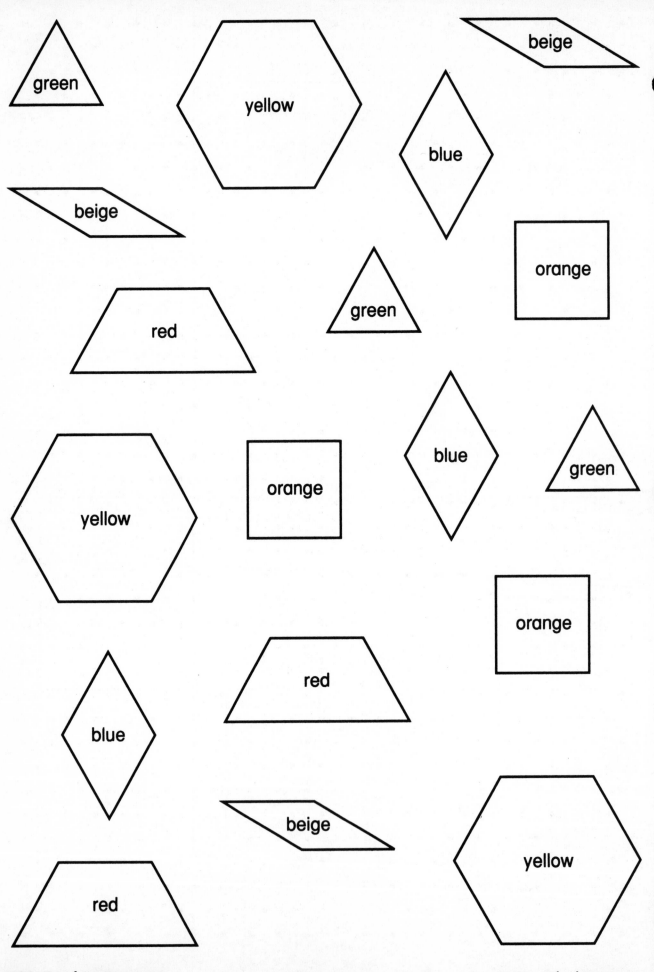

Pattern Block Patterns

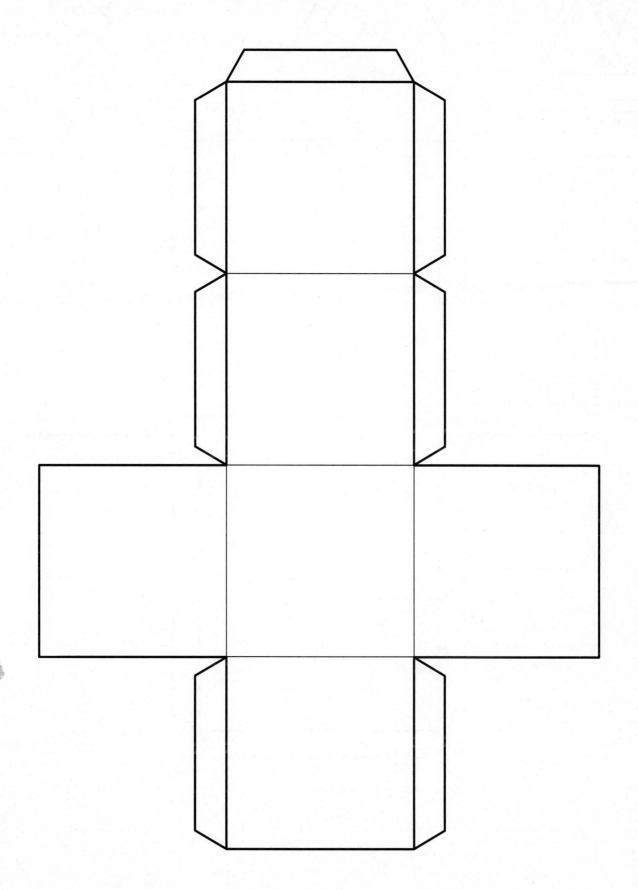

Cube Pattern

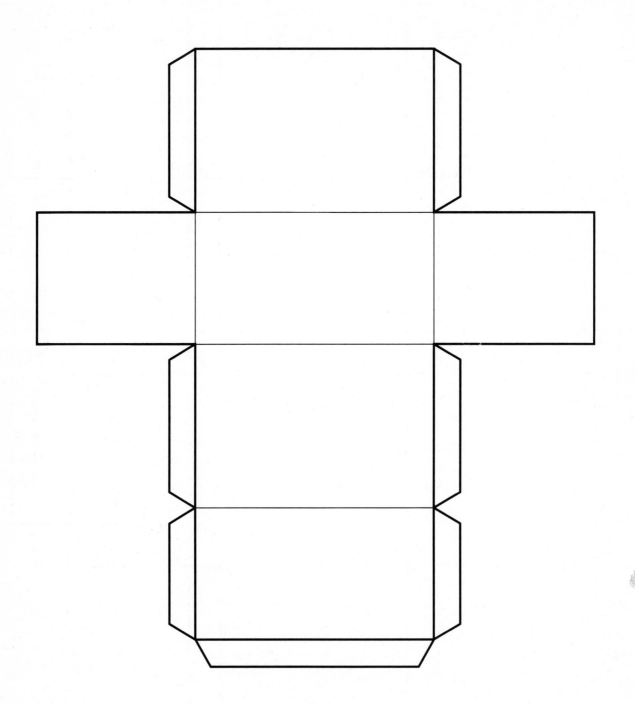

Rectangular Prism Pattern

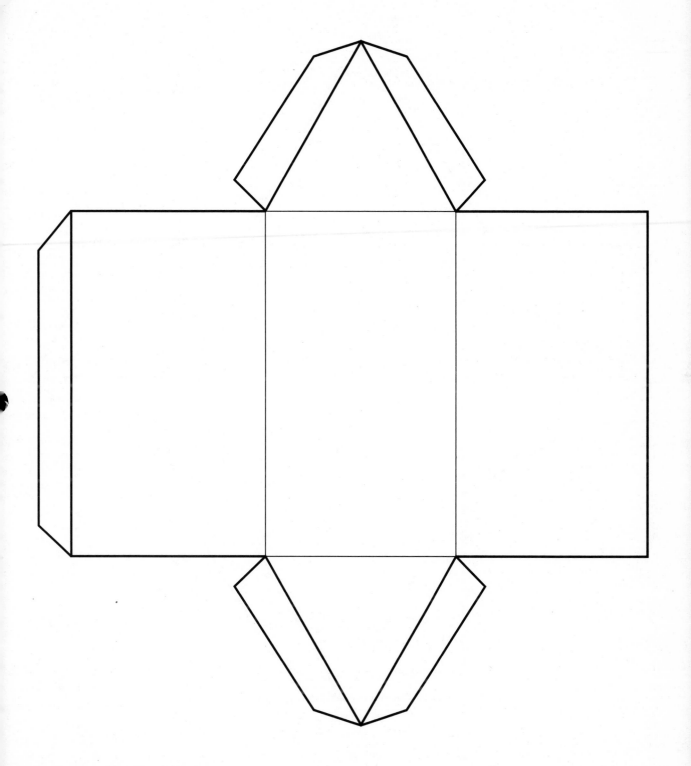

Triangular Prism Pattern

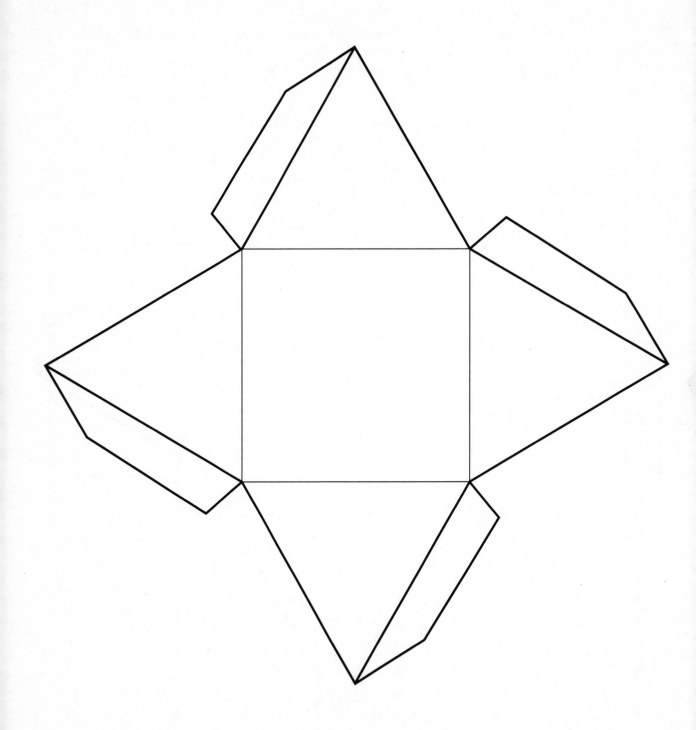

Square Pyramid Pattern

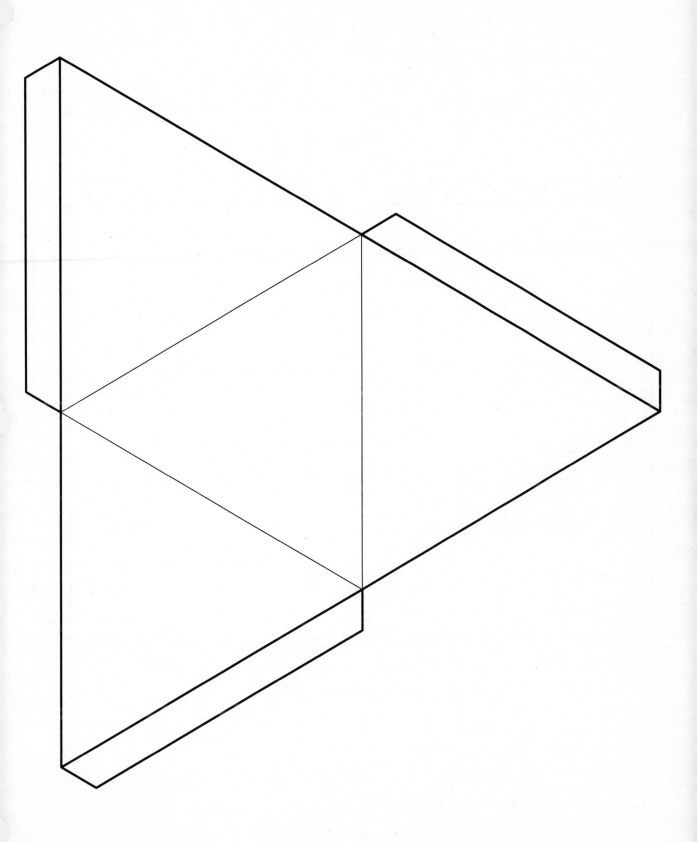

Triangular Pyramid Pattern

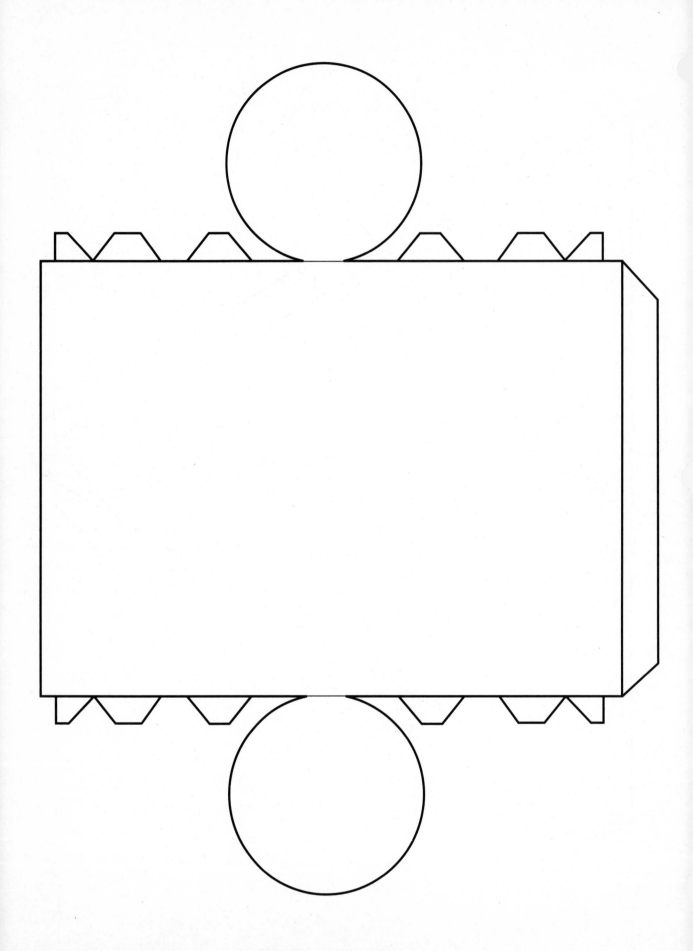

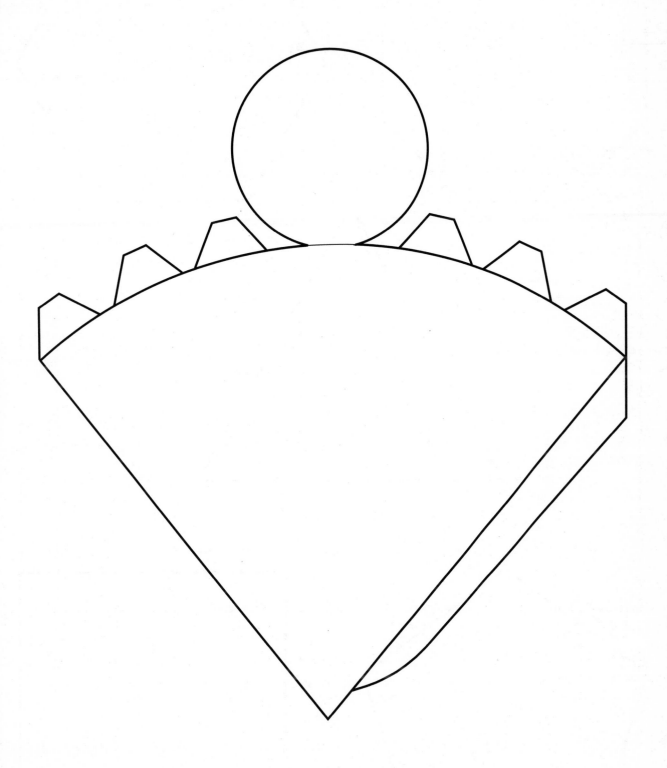

Cone Pattern

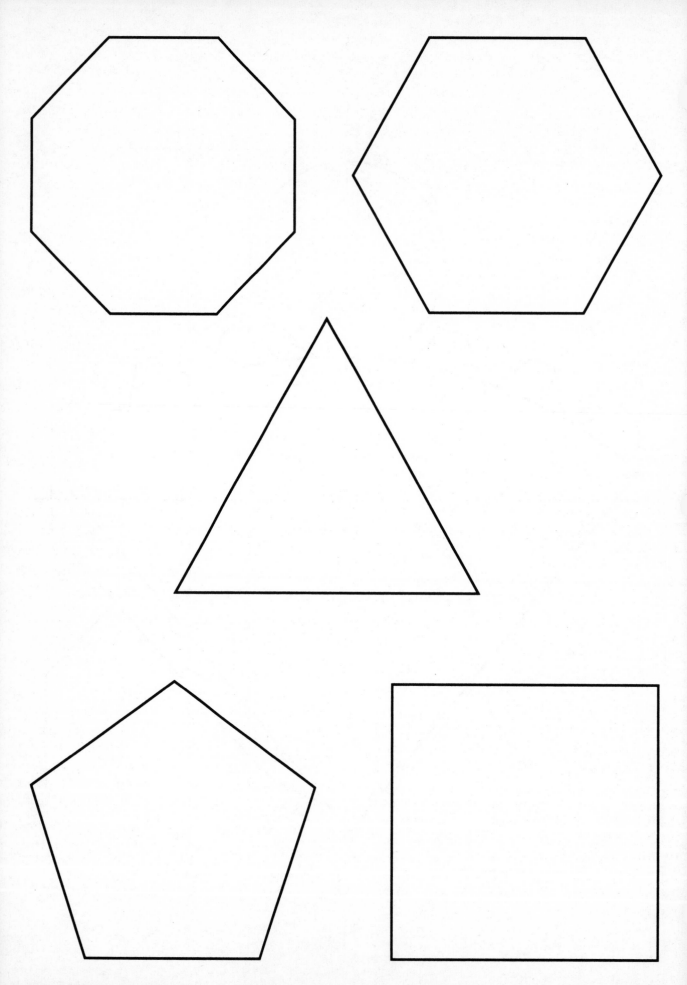

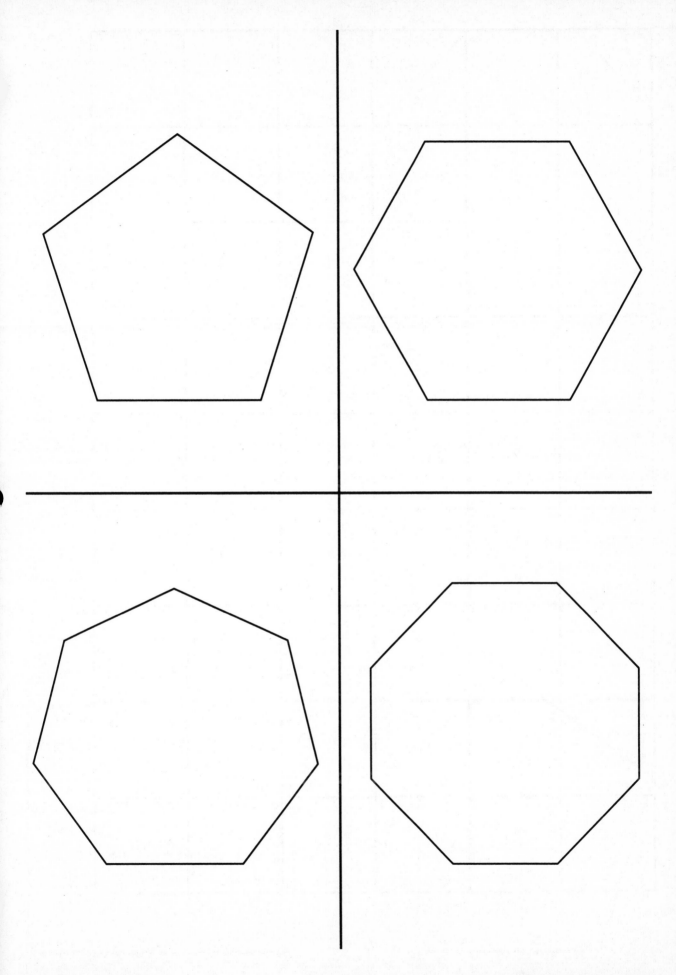

Polygons: Multi-sided

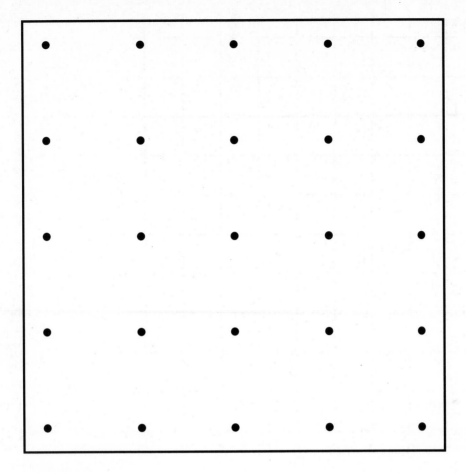

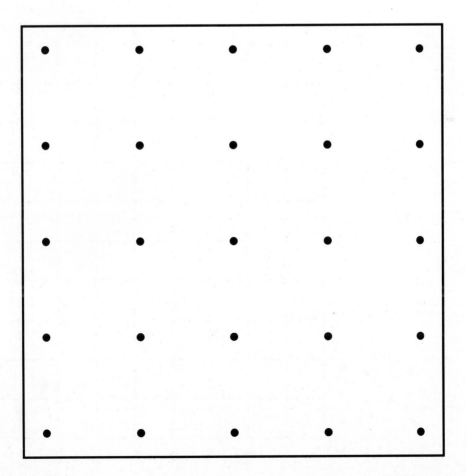

Geoboard Dot Paper

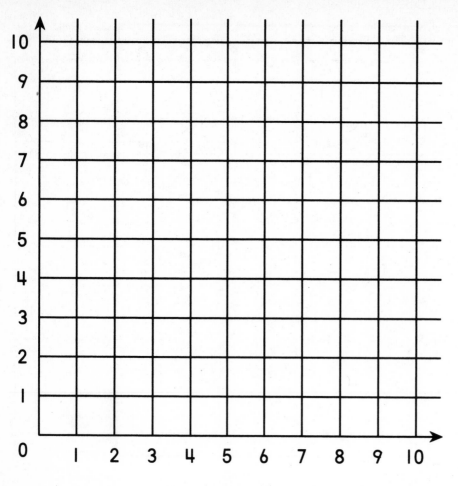

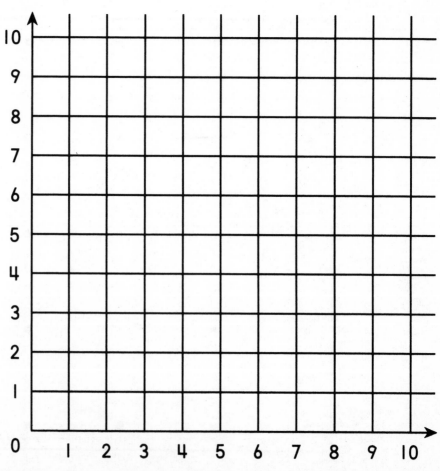

Grid of Quadrant I

Title _____

0

1-Centimeter Grid Paper

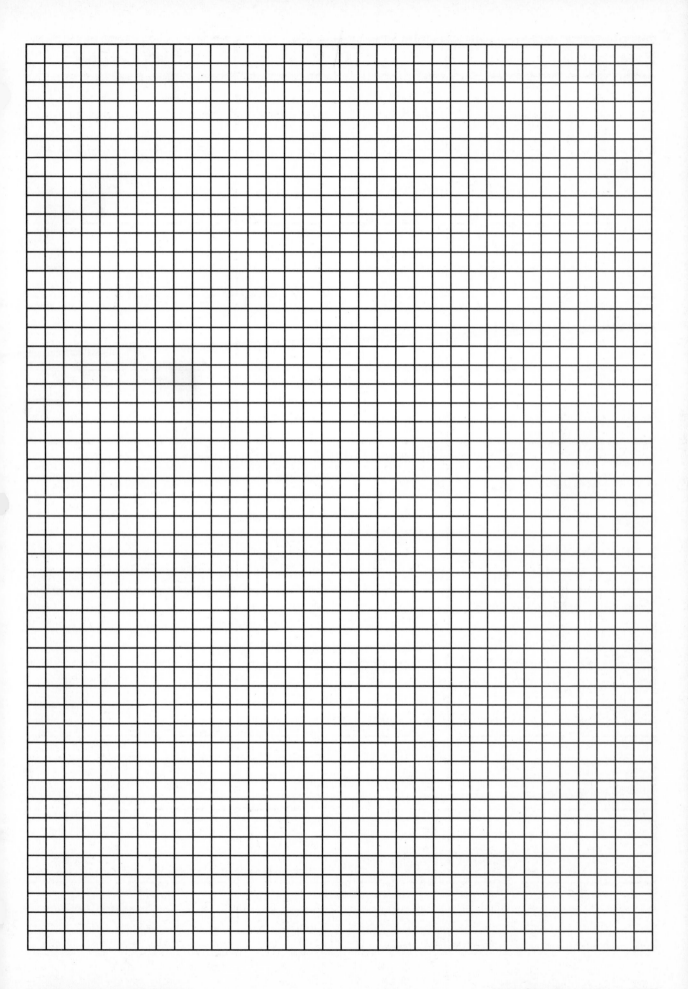

0.5-Centimeter Grid Paper

	Tally	Frequency

Tally Table

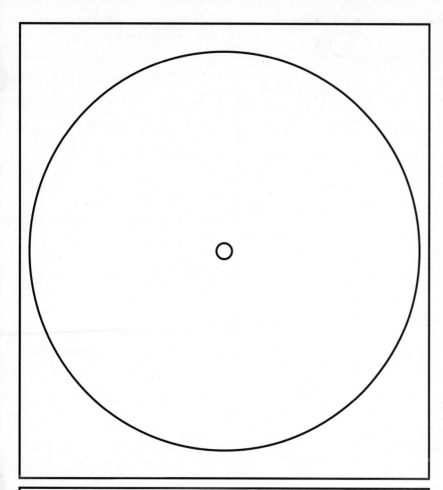

Spinner Tips

How to assemble spinner.
- Glue patterns to oaktag.
- Cut out and attach pointer with a fastener.

Alternative
- Students can use a paper clip and pencil instead.

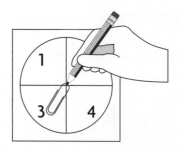

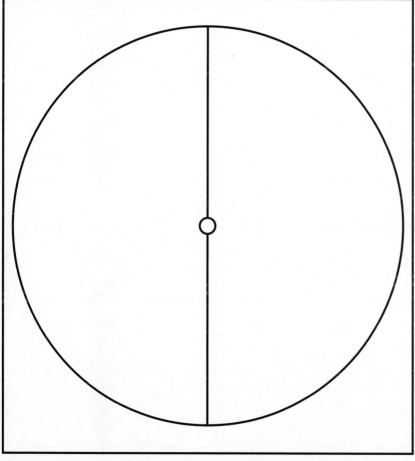

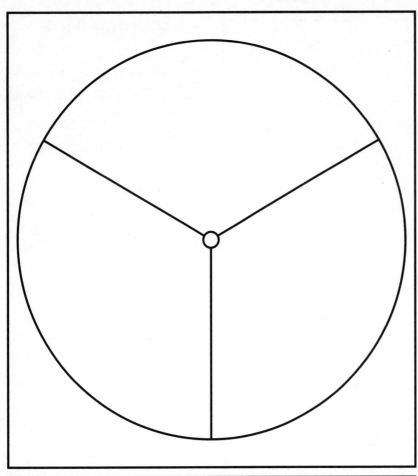

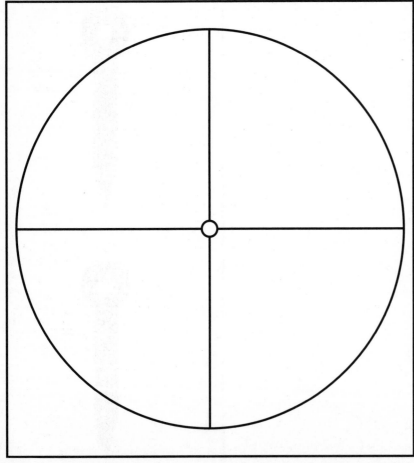

Spinner Tips

How to assemble spinner.
- Glue patterns to oaktag.
- Cut out and attach pointer with a fastener.

Alternative
- Students can use a paper clip and pencil instead.

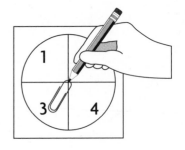

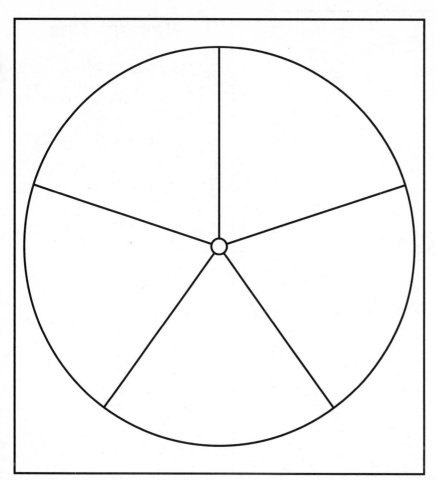

Spinner Tips

How to assemble spinner.
- Glue patterns to oaktag.
- Cut out and attach pointer with a fastener.

Alternative
- Students can use a paper clip and pencil instead.

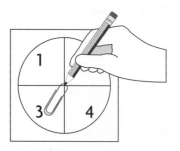

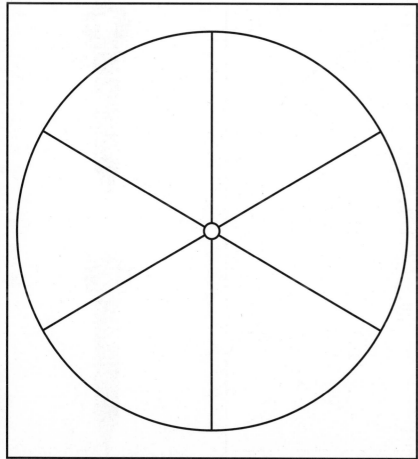

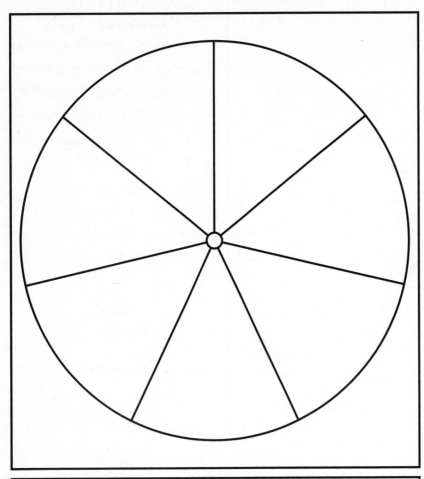

Spinner Tips

How to assemble spinner.
- Glue patterns to oaktag.
- Cut out and attach pointer with a fastener.

Alternative
- Students can use a paper clip and pencil instead.

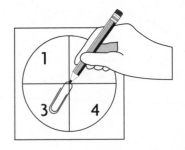

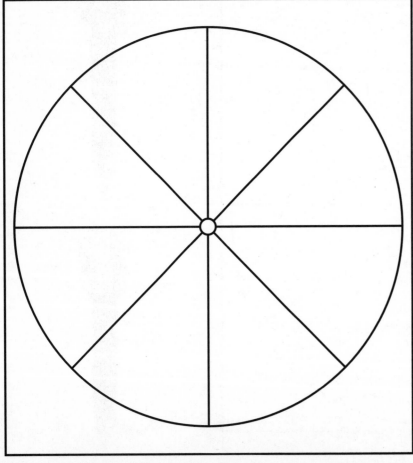

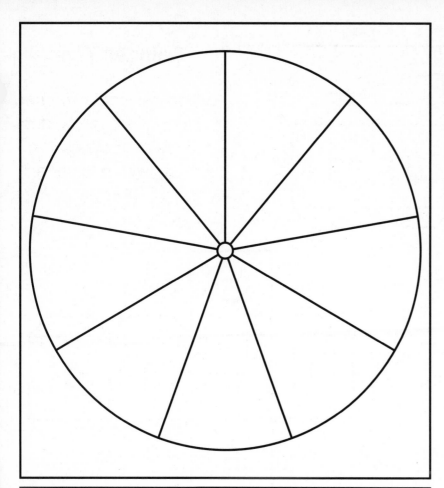

Spinner Tips

How to assemble spinner.
- Glue patterns to oaktag.
- Cut out and attach pointer with a fastener.

Alternative
- Students can use a paper clip and pencil instead.

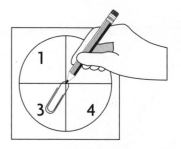

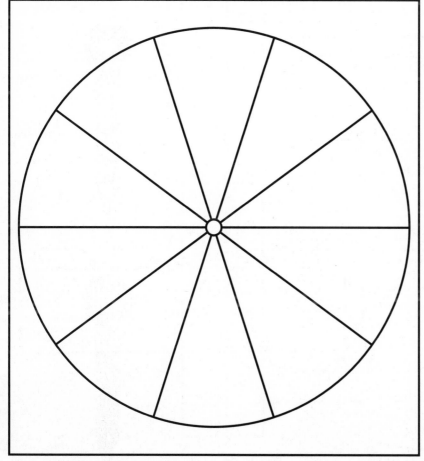

Spinners (9- and 10-section)

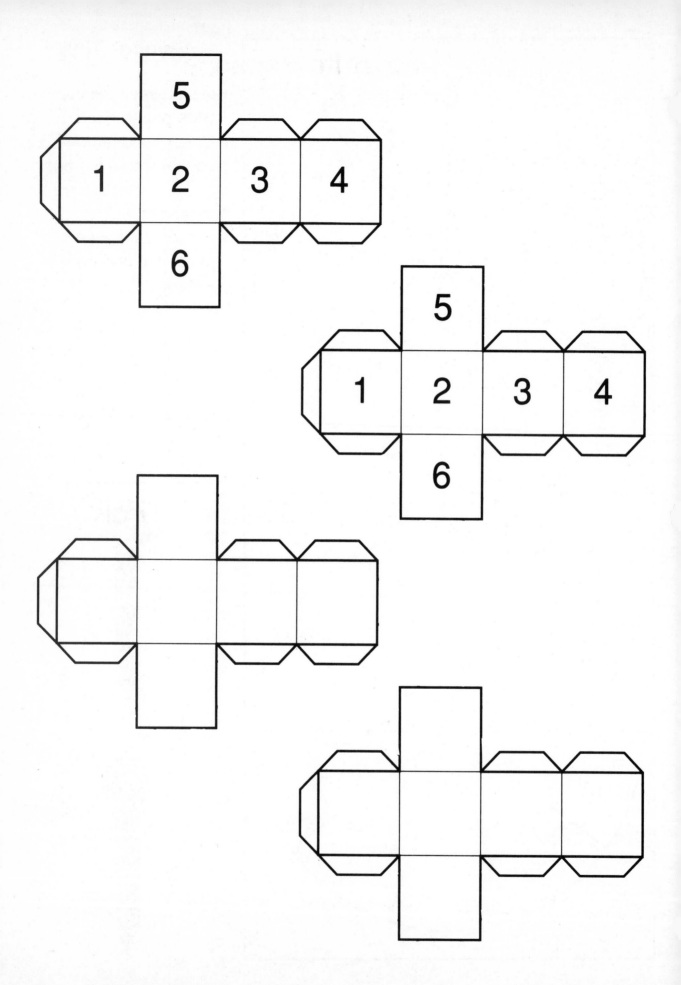

Favorite Seasons
Chosen by 16 Students

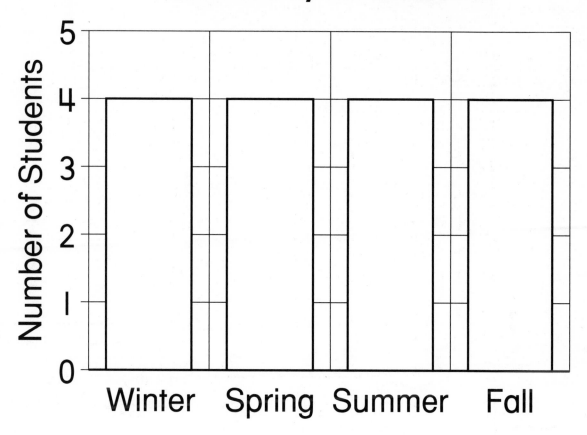

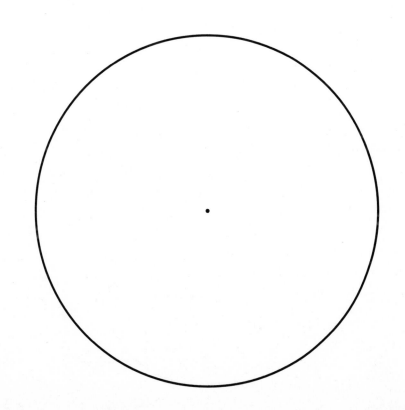

The 16 Students on the Yearbook Staff

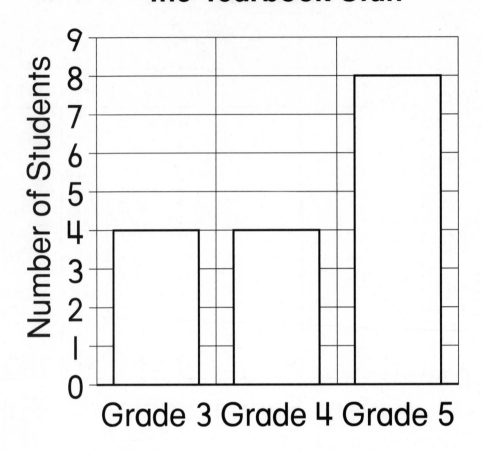

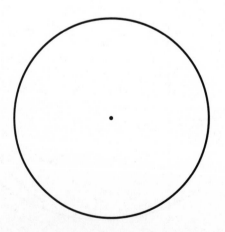

Favorite Juices
of 18 Students

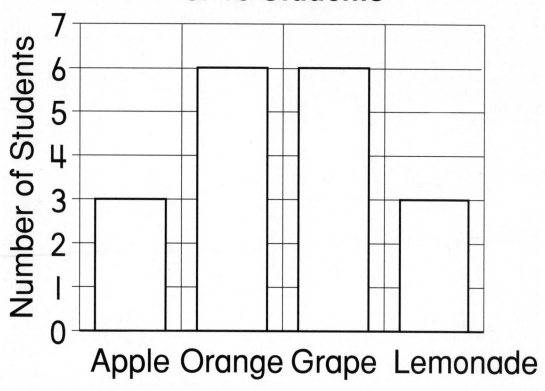

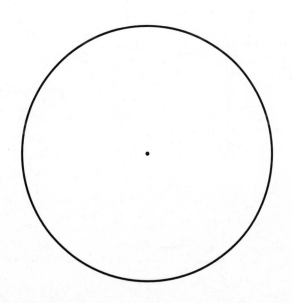

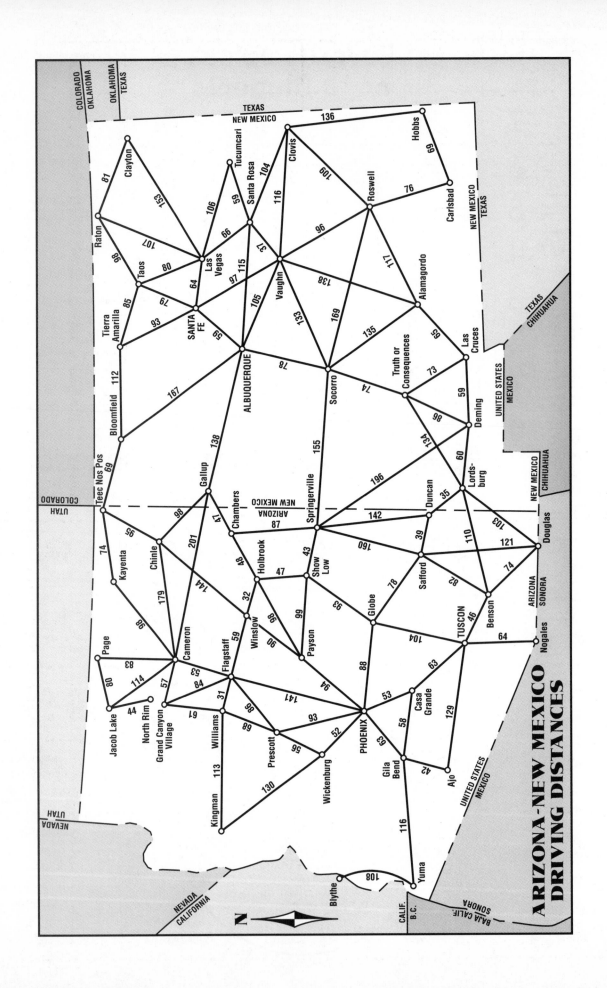

ARIZONA-NEW MEXICO
DRIVING DISTANCES

2	1	7	3	5	8
4	3	5	2	4	6
0	7	3	1	9	4
7	2	9	5	4	7
8	4	6	8	1	5
2	4	0	7	6	3
8	2	5	6	4	9

0	4	6	30	15
0	64	18	45	8
20	12	4	30	24
81	10	56	27	49
15	64	30	15	18
45	8	18	20	12
24	48	72	32	45

+	1	2	3	4	5	6	7	8	9
1	1	2	3	4	5	6	7	8	9
2	2	4	6	8	10	12	14	16	18
3	3	6	9	12	15	18	21	24	27
4	4	8	12	16	20	24	28	32	36
5	5	10	15	20	25	30	35	40	45
6	6	12	18	24	30	36	42	48	54
7	7	14	21	28	35	42	49	56	63
8	8	16	24	32	40	48	56	64	72
9	9	18	27	36	45	54	63	72	81

For use with Chapter 4 Practice Game

COMPARE TWO

START			
901 **910**	Lose 1 turn.	**89** **67**	200 600
862 **866**	299 266		
Toss Again.	5,432 4,321		
225 **265**	10 101	*1,991* *1,091*	END
Go Ahead 2.			
5,000 9,000	*123* *213*	59 53	**162** **165**

8,900 9,800

2,104 2,401

Go Back 2.

599 549

230 330

Toss Again.

Put in order (least to greatest)

Put in order (least to greatest)

Put in order (greatest to least)

Put in order (greatest to least)

For use with Chapter 7 Practice Game

Which kind of graph would best show the cost of a loaf of bread over the past 25 years? line graph	Which kind of graph would best show the field-trip choices of fourth graders? pictograph or bar graph	Which kind of graph would best compare two groups, such as a group of boys and a group of girls? double bar graph
Which kind of graph would best show how many students were absent each month during the school year? line graph	Which kind of graph would best show the median age of teachers in your school? stem-and-leaf plot	Which kind of graph would best show the number of pages each of several students could read in an hour? bar graph
Which kind of graph would best show the means of transportation each student uses for getting to school? pictograph or bar graph	Which kind of graph would best show the types of music enjoyed by students in a fourth-grade class? pictograph or bar graph	Which kind of graph would best compare the favorite sports of two classes? double bar graph
Which kind of graph would best show the median number of sit-ups for students in your class? stem-and-leaf plot	Which kind of graph would best show the growth of a baby during the first year of life? line graph	Which kind of graph would best show the number of fourth graders who had visited famous places? pictograph or bar graph
Which kind of graph would best show the amount of rainfall over the past year? line graph	Which kind of graph would best show the median grade students in your class received on the math quiz? stem-and-leaf plot	Which kind of graph would best show the favorite authors of students in a fourth-grade class? pictograph or bar graph

Graph Score Sheet

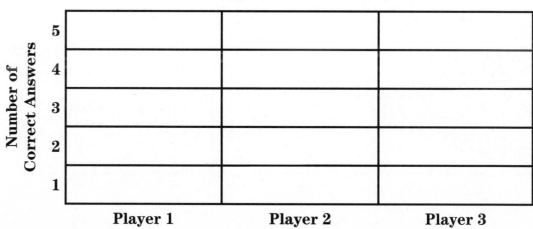

Angles from A to . . .

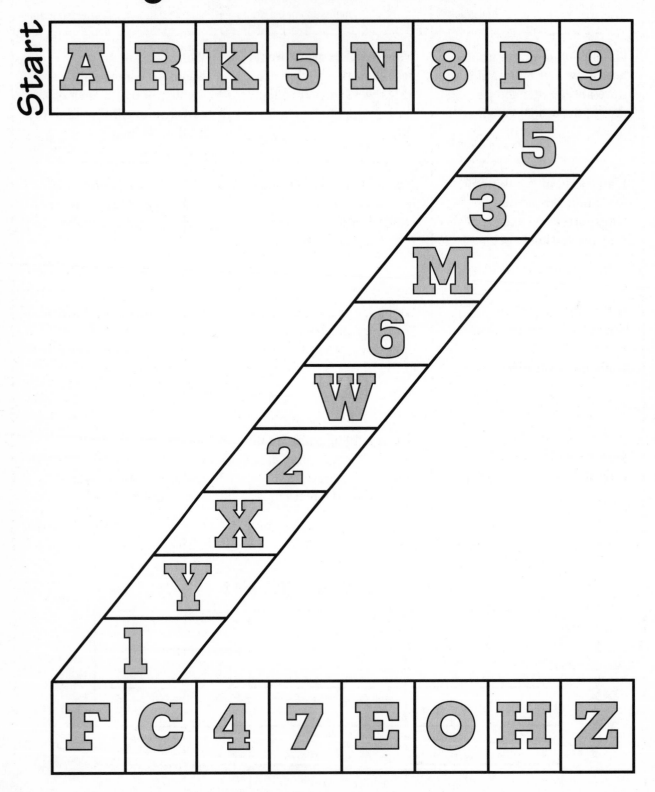

For use with Chapter 13 Practice Game

For use with Chapter 15 Practice Game

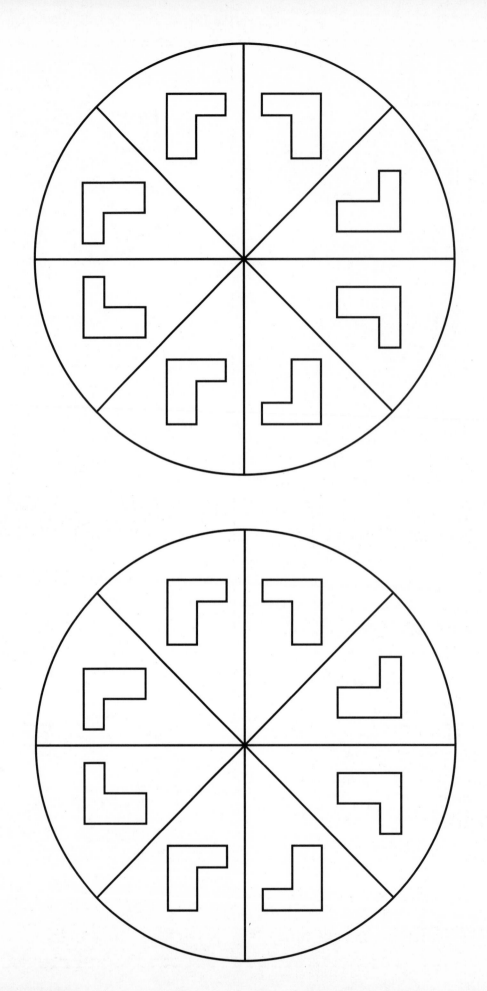

Television

Shirt

Audio Cassette Tape

Skates

Portable Stereo

VCR

2	3
4	5
10	20
30	40
50	60

0.1	0.3	0.5	0.6	0.7
$\dfrac{1}{10}$	$\dfrac{3}{10}$	$\dfrac{5}{10}$	$\dfrac{6}{10}$	$\dfrac{7}{10}$
0.10	0.15	0.30	0.35	0.51
$\dfrac{10}{100}$	$\dfrac{15}{100}$	$\dfrac{30}{100}$	$\dfrac{35}{100}$	$\dfrac{51}{100}$
0.64	0.70	0.77	0.82	0.89
$\dfrac{64}{100}$	$\dfrac{70}{100}$	$\dfrac{77}{100}$	$\dfrac{82}{100}$	$\dfrac{89}{100}$
1.5	2.3	3.6	7.12	8.22
$1\dfrac{5}{10}$	$2\dfrac{3}{10}$	$3\dfrac{6}{10}$	$7\dfrac{12}{10}$	$8\dfrac{22}{100}$

For use with Chapter 22 Practice Game

0.01	0.02	0.03	0.04	0.05
0.05	0.05	0.06	0.07	0.08
0.09	0.10	0.10	0.11	0.12
0.13	0.14	0.15	0.16	0.17
0.18	0.19	0.20	0.25	0.30

For use with Chapter 23 Practice Game

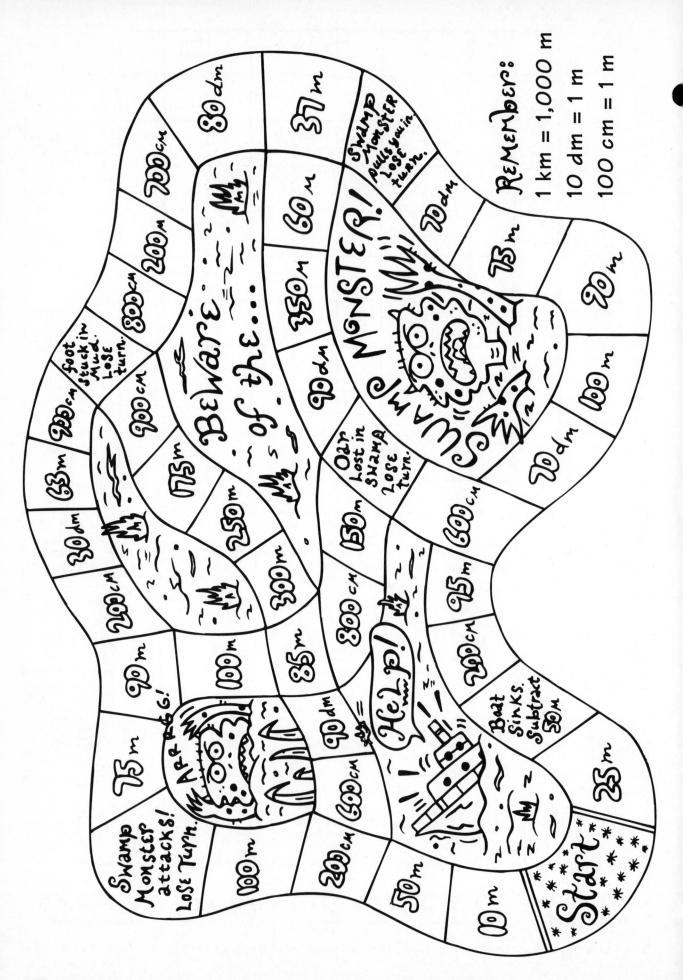

For use with Chapter 25 Practice Game

HOT SPOTS

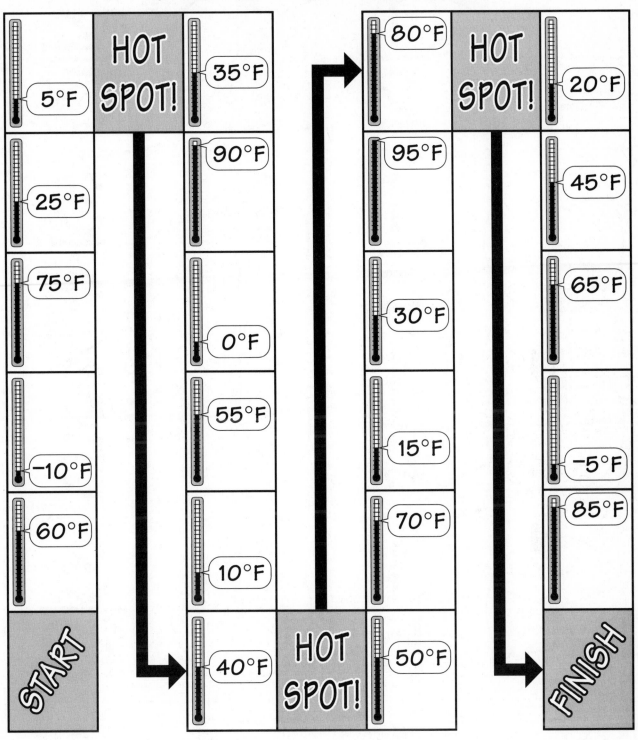

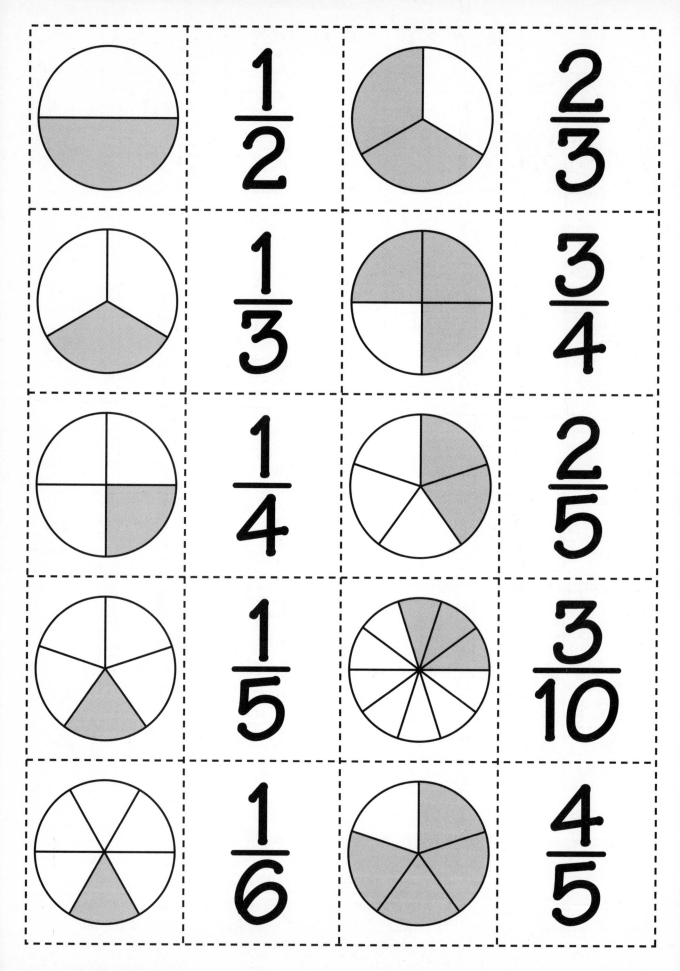

For use with Chapter 28 Practice Game

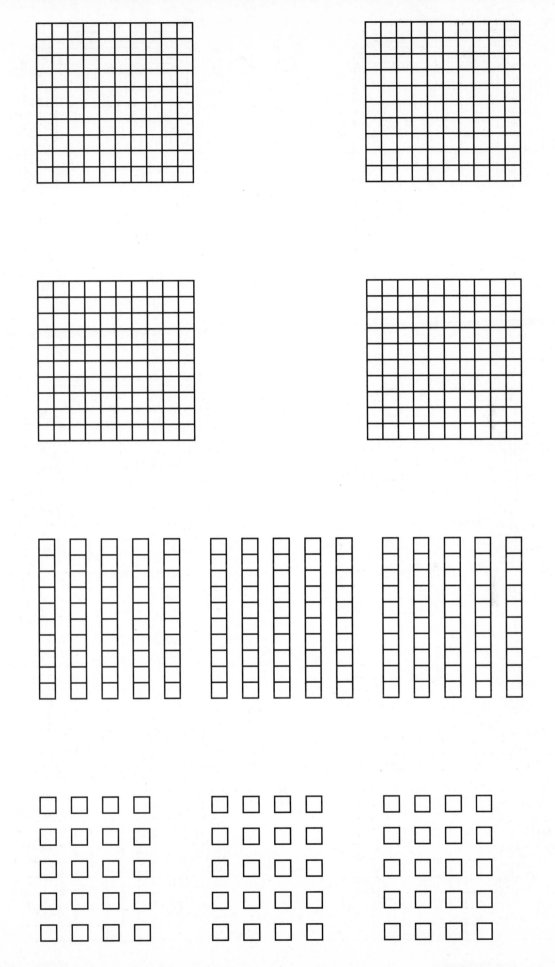

For use with Practice Activities 5A and 5B

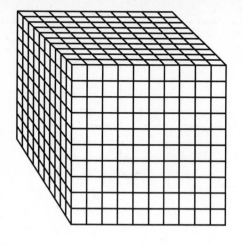

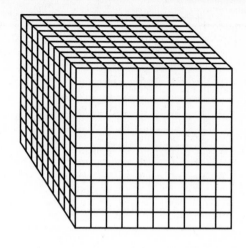

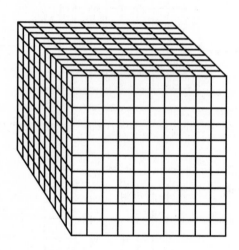

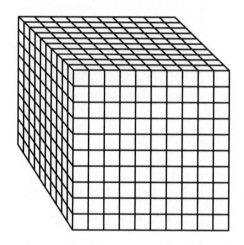

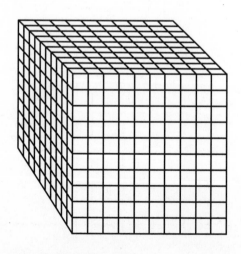

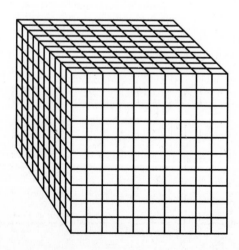

Thousands	Hundreds	Tens	Ones

Ten Thousands	Thousands	Hundreds	Tens	Ones

For use with Practice Activities 5A and 5B

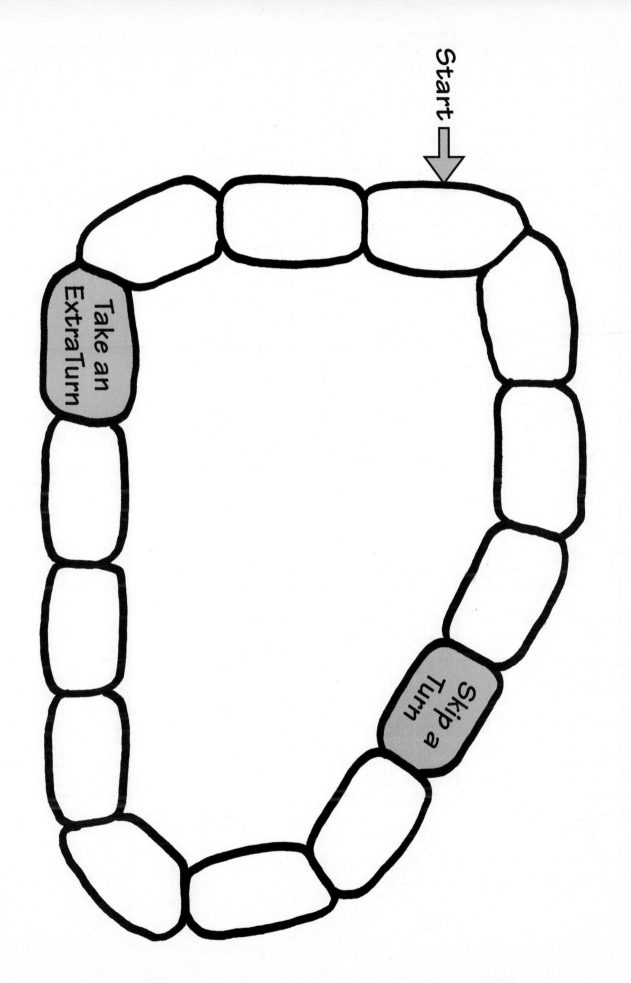

Start

Take an Extra Turn

Skip a Turn

For use with Practice Activities 21A, 21B, 23A, and 23B

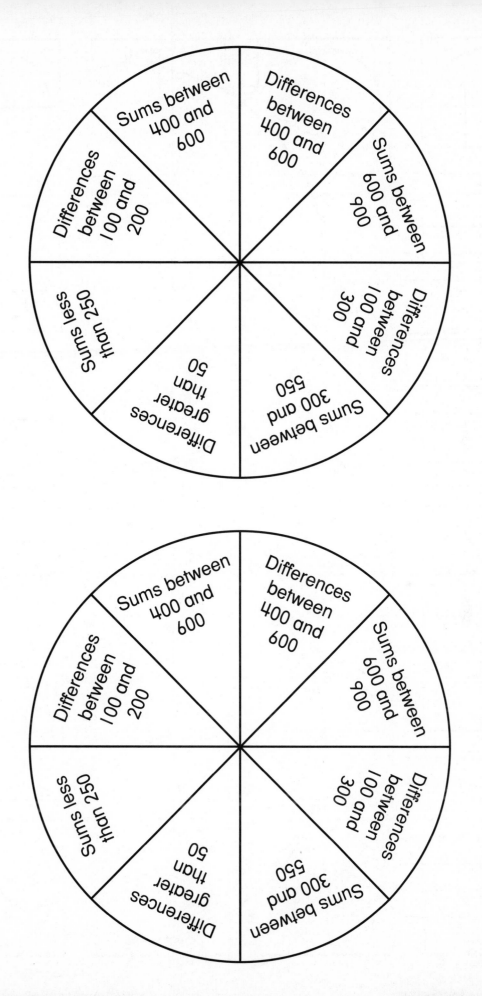

Multiply by _____

Factor	Product
0	
1	
2	
3	
4	
5	
6	
7	
8	
9	
10	

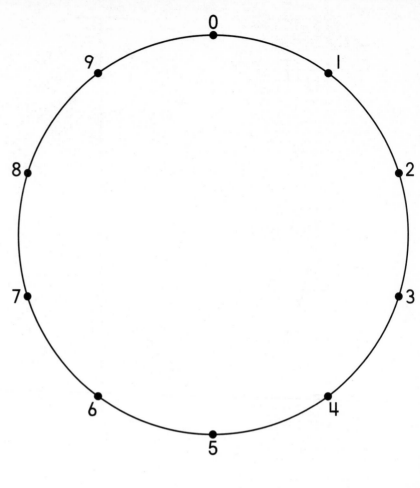

Multiply by _____

Factor	Product
0	
1	
2	
3	
4	
5	
6	
7	
8	
9	
10	

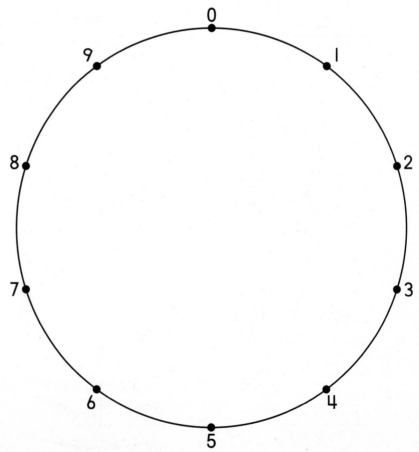

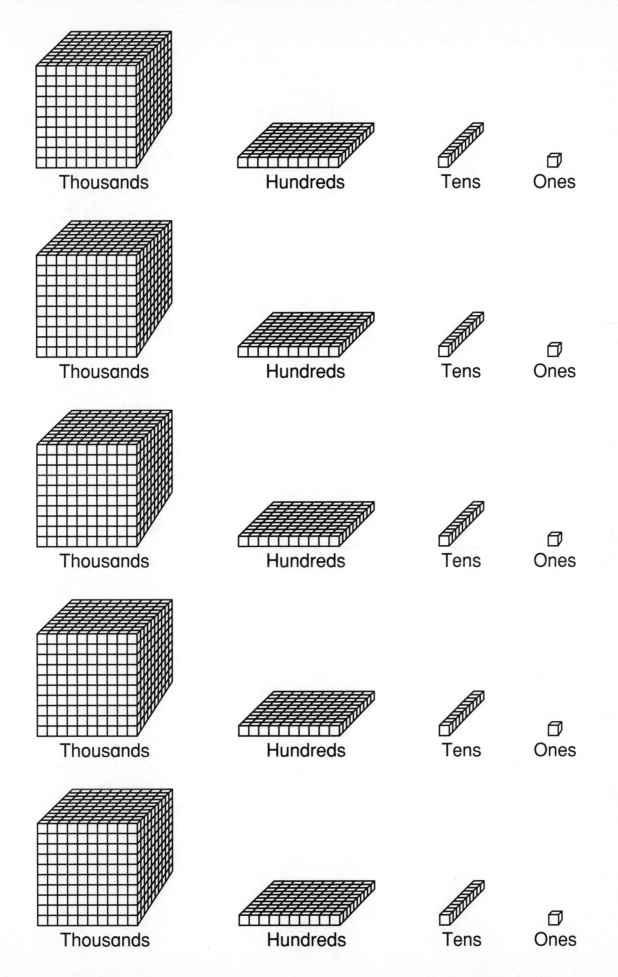

Thousands　　　Hundreds　　　Tens　　　Ones

Thousands　　　Hundreds　　　Tens　　　Ones

Thousands　　　Hundreds　　　Tens　　　Ones

Thousands　　　Hundreds　　　Tens　　　Ones

Thousands　　　Hundreds　　　Tens　　　Ones

						Hundred Thousands
						Ten Thousands
						Thousands
						Hundreds
						Tens
						Ones

Prediction for 10 spins:

	Color	Tally	Frequency
Results:	Red		
	Green		
	Yellow		

Prediction for 50 spins:

	Color	Tally	Frequency
Results:	Red		
	Green		
	Yellow		

How do predictions and results compare?

A triangle with one right angle	A triangle with an acute angle	A figure with 5 vertices. Add a line segment to make a right triangle.
Four right triangles connected at the center of the geoboard	A square, a parallel segment, and a line segment that intersects both	A hexagon
A triangle with one obtuse angle	Two perpendicular segments and a rectangle with one side made from each segment	A hexagon with two perpendicular segments inside; add a segment to make a right triangle.
A rectangle and two triangles with obtuse angles that cross over each other	A figure with 5 vertices	Two line segments that go from corner to corner and only intersect at their center. Add an inner square and an outer square.

Use the digits 1, 2, 3, 4 to get the smallest possible product.

Use the digits 1, 2, 3, 4 to get the greatest possible product.

Use the digits 2, 4, 6, 8 to get the smallest possible product.

Use the digits 2, 4, 6, 8 to get the greatest possible product.

Use the digits 1, 3, 5, 7 to get the greatest possible product.

Use the digits 6, 3, 0, 5 to get the greatest possible product.

Use the digits 1, 6, 7, 9 to get the greatest possible product.

Use the digits 9, 8, 1, 3 to get the smallest possible product.

Use the digits 2, 4, 6, 8 to get the product closest to but not greater than 2,000.

Use the digits 2, 4, 6, 8 to get the product closest to but not greater than 3,500.

Use the digits 6, 7, 8, 9 to get the product closest to but not greater than 4,500.

Use the digits 4, 5, 6, 9 to get the product closest to but not greater than 5,000.

$$\frac{3}{5} + \frac{1}{5} = \boxed{}$$

$$\frac{1}{3} + \frac{1}{3} = \boxed{}$$

$$\frac{2}{2} - \frac{1}{2} = \boxed{}$$

$$\frac{7}{10} - \frac{5}{10} = \boxed{}$$

$$\frac{7}{10} + \frac{2}{10} = \boxed{}$$

$$\frac{3}{12} + \frac{3}{12} = \boxed{}$$

$$\frac{9}{12} - \frac{6}{12} = \boxed{}$$

$$\frac{4}{4} - \boxed{} = \frac{3}{4}$$

$$\frac{2}{8} + \frac{4}{8} = \boxed{}$$

$$\frac{3}{6} + \boxed{} = \frac{6}{6}$$

$$\frac{8}{10} - \frac{3}{10} = \boxed{}$$

$$\frac{7}{8} - \boxed{} = \frac{3}{8}$$

$$\frac{1}{8} + \frac{4}{8} = \boxed{}$$

$$\frac{1}{4} + \boxed{} = \frac{2}{4}$$

$$\frac{5}{6} - \frac{3}{6} = \boxed{}$$

$$\frac{6}{7} - \frac{2}{7} = \boxed{}$$

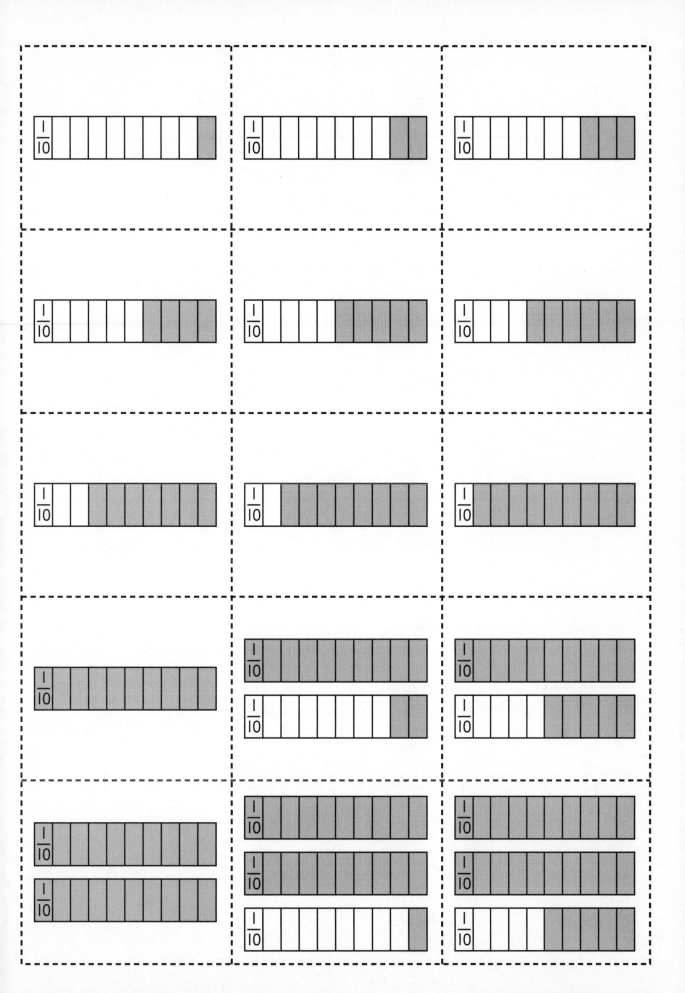

$\dfrac{1}{10}$	$\dfrac{2}{10}$	$\dfrac{3}{10}$
$\dfrac{4}{10}$	$\dfrac{5}{10}$	$\dfrac{6}{10}$
$\dfrac{7}{10}$	$\dfrac{8}{10}$	$\dfrac{9}{10}$
1	$1\dfrac{2}{10}$	$1\dfrac{5}{10}$
2	$2\dfrac{1}{10}$	$2\dfrac{5}{10}$

For use with Learning Center Card 22

0.1	0.2	0.3
0.4	0.5	0.6
0.7	0.8	0.9
1.0	1.2	1.5
2.0	2.1	2.5

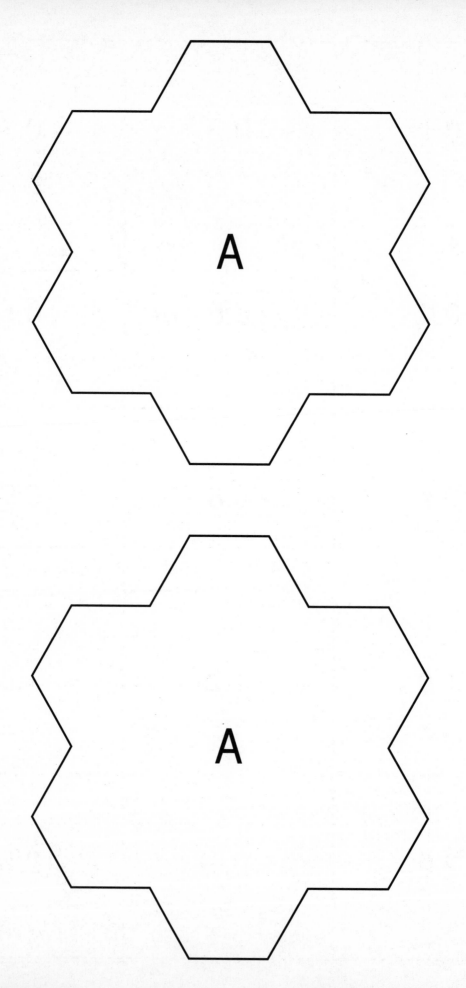